THINK LIKE A SCIENTIST™

WHAT DO YOU WANT TO PROVE?

Planning Investigations

Barbara A. Somervill

The Rosen Publishing Group's
PowerKids Press™
New York

For Chuck

Published in 2007 by The Rosen Publishing Group, Inc.
29 East 21st Street, New York, NY 10010

First Edition

Editor: Joanne Randolph
Book Design: Elana Davidian
Layout Design: Julio Gil

Photo Credits: Cover © Will & Deni McIntyre/Photo Researchers, Inc.; pp. 4, 12 © Richard Hutchings/Photo Researchers, Inc.; p. 7 © Cristina Pedrazzini/Photo Researchers, Inc.; p. 8 © David Noton/Masterfile; p. 11 © Bob Daemmrich/Photo Edit; pp. 15, 20 Cindy Reiman; p. 16 © Will Hart/Photo Edit; p. 19 © Charles D. Winters/Photo Researchers, Inc.

Library of Congress Cataloging-in-Publication Data

Somervill, Barbara A.
What do you want to prove? : planning investigations / Barbara A. Somervill.— 1st ed.
p. cm. — (Think like a scientist)
Includes index.
ISBN 1-4042-3483-7 (library binding) — ISBN 1-4042-2192-1 (pbk.)
1. Science—Experiments—Juvenile literature. 2. Science—Methodology—Juvenile literature. I. Title. II. Series.
Q164.S612 2007
507.8—dc22

2005030265

Manufactured in the United States of America

Contents

Observing, or looking closely, at what happens during a science experiment is part of the scientific method.

The Scientific Method

Scientists ask questions every day. Then they use the **scientific method** to find answers. The first step in the scientific method is to form a **hypothesis**, or a good guess about the question's answer. Next the scientist plans an experiment to prove the hypothesis.

Here's an example. Many people ask how they can reduce water waste. A scientist wonders if **recycled** water from washing clothes would successfully help grow grass. She guesses it will not. The effect of using recycled water on grass can be tested using an experiment. To get an exact answer, an experiment is done several times. Each time is called a run, or trial. The scientist compares the results of each trial to discover if the hypothesis was right.

What Are You Trying to Prove?

The first step in planning an **investigation** is knowing what you are trying to prove. Keep your focus, or central idea, narrow and your experiment simple. It is easier to prove one small idea than several large ideas.

For example, an experiment on the effect of using recycled water to grow grass is easy to do. Showing that recycled water can be used for all plants is too hard. Such an experiment would mean growing thousands of plants.

Your plan must consider space, cost, and time. Plan your experiment to fit the amount of room you have to use. Find out what your materials will cost. Materials are the items you need for the experiment. Decide how much time you can spend on the experiment. These elements will help you keep your experiment focused.

There are thousands of different kinds of grass. As part of your plan for your experiment, you should choose the kind of grass you want to use.

It is helpful to learn all you can about your subject. For this experiment you might ask a gardener, such as this person, which kinds of grass are easiest to grow.

Setting Up an Experiment

How do you plan an experiment? Your hypothesis says that you think recycled laundry rinse water will not help grass grow as well as regular tap water would. Now you must prove your hypothesis true or false.

First do some **research**. Ask experts, or people who know a lot about a subject, about growing grass. Grass experts might include gardeners or garden store workers. You might also ask a librarian for help. The more you know in advance, the easier it is to make a plan.

You need to grow grass using recycled rinse water. You might use rinse water **sources** with three different types of detergent, or soap. Another approach would be using water with regular detergent and with **bleach**-added detergent. You also need to grow some grass using regular water. This is called a control.

A Needed Control

Every experiment needs a control. This is an experiment that begins with normal conditions and produces an expected result. In the grass-growing experiment, the control is grass grown using tap water.

The control experiment becomes a measuring stick. It produces grass that grows at the expected or normal rate. You want to find out if the grass watered with recycled water grows better or worse than the control.

You must set up **constant** conditions for your grass experiment. Plant all the grass at the same time. Use the same soil, grass seed, light source, and **temperature**. Water all pots with the same amount of water, changing only the water source. The water sources are the experiment's **variables**.

In any experiment you do, you will want to have a control. This girl is testing what happens when she mixes certain liquids.

This scientist makes sure she changes only one variable at a time while she runs her experiment. She needs to be able to figure out what caused certain results.

What Are Variables?

The things that change in an experiment are called variables. In the grass-growing experiment, the variable is the water source. In a good experiment, a scientist changes only one thing at a time.

Suppose you used different types of seed and soil and planted each pot at a different time. You would have too many variables. Did the grass in pot *A* grow better because of the seed, soil, time, or water source?

You must even control the use of each water source. Give each pot the same amount of water, about ¼ cup every other day, no matter which source of water you are using. By limiting the variables, you can decide if the variable you changed caused the results you see.

A Materials List

A science experiment is like baking. You use certain **ingredients** and supplies to produce a good result. In an experiment the materials list gives the same **information**.

Your list must be exact. It must include all the materials used and how much of each is used. An experiment to grow grass requires soil, grass seed, pots, and water sources. First you must plan how many pots of grass you want to grow. Then you will need to figure out how much of each supply you need. You can do this by finding out how much you need for one pot and multiplying it by the total number of pots. Planning an investigation means you need to think through each step of your experiment.

AMOUNT	MATERIAL
6	3-inch (8 cm) plastic pots with holes in the bottom
6 tablespoons	fescue or other grass seed
12 cups	potting soil

AMOUNT	MATERIAL
4 cups	tap water
4 cups	rinse water with detergent
4 cups	rinse water with detergent with bleach added

List your supplies in a chart like this one. The amount of each item is on the left. The material you used is on the right. In this experiment you need pots, soil, measuring cups and spoons, and a ruler, as shown here.

Taking good notes about your plan will help you repeat your experiment more than once and make sense of your results.

Getting Started

To start your experiment, you must make a step-by-step plan. What do you do first? What comes next? You must carefully describe every step you take.

You have made your materials and equipment lists. That was the first step toward starting your experiment. Next you must tell how to use the materials. Give exact directions. Do not write, "Put soil in pots." Write, "Put 2 cups of soil in each 3-inch (8 cm) pot."

Write down the steps you plan to take to do your experiment. Leave space between each step in case you forget something. You can fill in any extra steps as you do your experiment the first time. After you finish setting up, number each step. If you write step-by-step directions, you or another scientist can follow them again to repeat the experiment.

You will need to measure the results of your experiment. Think about this as you plan. What tools will you need? You might need to measure height, weight, volume, or time. Common measurement tools include rulers, scales, and stopwatches.

Date and record each measurement in a chart or table. This lets you see how your measurements have changed. You can use a table to compare results in different parts of an experiment.

It is best to use the **metric system** for measuring your results, because it is more exact. For your grass experiment, use a metric ruler to measure centimeters or millimeters. You must use the same measurement tools every time you measure. These tools help you get exact results.

Scientists use glass tubes with markings on them to help them measure an exact amount of whatever they are testing. These tubes are called beakers, test tubes, or graduated cylinders.

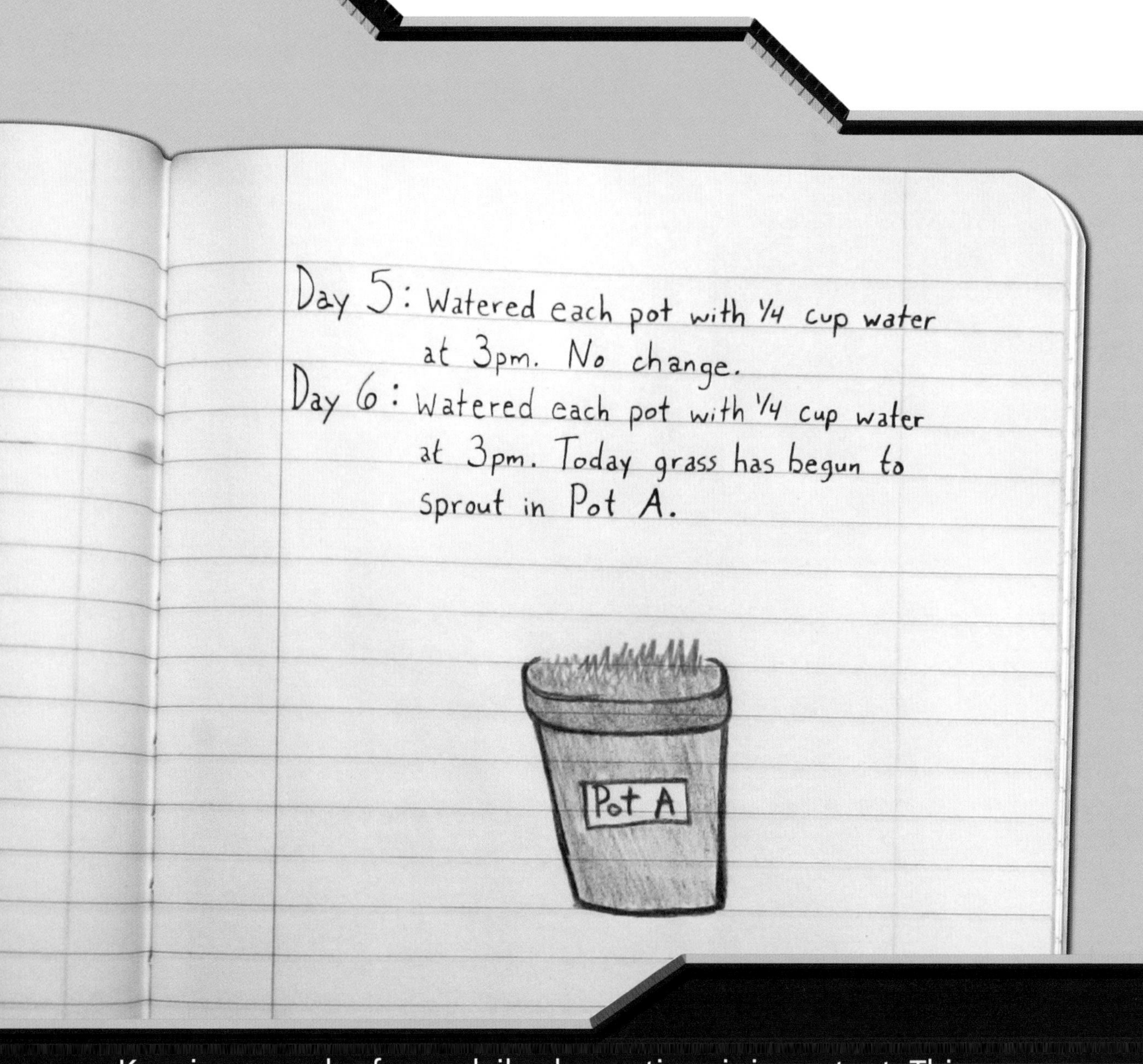

Keeping records of your daily observations is important. This way you will know exactly what happened when you look back through your notes. Drawings can be a helpful way to record information, too.

Keeping a Record

Scientists collect information as they run their experiments. They make **observations**. They measure and record changes. Recording information takes place at regular periods. You must record information often, such as every day or three times a week. Build this into your experiment plan.

An observation is noting a change that is seen, heard, felt, smelled, or tasted. When you notice any change, write it down. For example, if grass seed sprouts in pot *A* first, record it. It may be just green fuzz, but it is something new.

As you keep records of what you see and measure, mark the most interesting events with a highlighter. The day the grass sprouts is a highlight event. You could record the event with a photo or a diagram.

What Did You Learn?

Several key elements are important when planning an investigation. You need to make step-by-step plans. A good plan has plenty of information.

When you start your experiment, you should leave yourself enough time to complete two or three trials. Then you can compare the results of each trial. The results should be the same. If they are not, what caused the difference? Did you follow the same **procedures** exactly?

If the rinse water with bleach produced the greatest grass growth, then the experiment's hypothesis was wrong. Why might bleach have increased grass growth? Now you have an idea for your next scientific investigation.

Glossary

bleach (BLEECH) A powder or liquid that makes things whiter or removes stains.

constant (KON-stent) Having to do with an element in an experiment that does not change.

hypothesis (hy-PAH-theh-ses) A possible answer to a problem.

information (in-fer-MAY-shun) Knowledge or facts.

ingredients (in-GREE-dee-unts) Any of the parts that go into a combination.

investigation (in-ves-tih-GAY-shun) The act of looking carefully in order to find facts.

metric system (MEH-trik SIS-tem) A method of measurement based on counting tens.

observations (ahb-ser-VAY-shunz) Things that are seen or noticed.

procedures (pruh-SEE-jurz) Steps or rules to follow.

recycled (ree-SY-kuld) Reused.

research (REE-surch) Careful study.

scientific method (sy-un-TIH-fik MEH-thid) The system of running experiments in science.

sources (SORS-ez) Things that give facts or knowledge.

temperature (TEM-pur-cher) How hot or cold something is.

variables (VER-ee-uh-bulz) Elements in an experiment that may be changed or may be constant.

Index

Web Sites

Due to the changing nature of Internet links, PowerKids Press has developed an online list of Web sites related to the subject of this book. This site is updated regularly. Please use this link to access the list:
www.powerkidslinks.com/usi/planinv/